E. Sheppard

Plants, Shrubs

Fruit and Ornamental Trees, Roses, Etc.

E. Sheppard

Plants, Shrubs
Fruit and Ornamental Trees, Roses, Etc.

ISBN/EAN: 9783741180538

Manufactured in Europe, USA, Canada, Australia, Japa

Cover: Foto ©Andreas Hilbeck / pixelio.de

Manufactured and distributed by brebook publishing software
(www.brebook.com)

E. Sheppard

Plants, Shrubs

ATALOGUE

OF

. SHEPPARD & SONS,

WHOLESALE AND RETAIL DEALERS IN

ANTS, SHRUBS,

uit and Ornamental Trees, Roses, Etc.

⇒1885⇐

GREEN-HOUSES AND NURSERY:

No. 224 Fairmount St., Lowell, Mass.

VOX PRESS, LOWELL, MASS.

Please Read This before Ordering.

IN presenting this Catalogue to our patrons, we would beg leave to return our thanks for the liberal patronage of the past, and trust to be favored with a continuance of the same.

Purchasers unacquainted with the different varieties of plants, leaving the selection with us, — stating the sum they wish to expend, and the object they wish to effect, — may depend upon receiving the best sorts and best plants.

In affixing prices to the various classes and varieties, it is difficult to always give the exact price, as the sizes vary, so we give only a price that will insure a good, strong, healthy plant and a first-class variety. It is our aim to deal in only the very best varieties.

Orders. — Parties ordering will please use the "ORDER SHEET," which will greatly facilitate their execution. If mixed up in the body of the letter, mistakes are almost unavoidable.

Persons starting in business, or florists in want of stock, will find it to their advantage to call upon us.

Shipping. — All plants sent out by us will be carefully packed and delivered in Lowell, and shipped or forwarded according to directions, after which they are at the risk of the purchaser.

We have for sale, this season, an unusually large and complete stock of Green-house and Bedding Plants, Evergreens, Shrubs, Vines, etc. We would earnestly request a personal inspection of the same. Those, however, favoring us with orders, may fully rely on our using every exertion to give perfect satisfaction.

Orders by mail or express will receive prompt and careful attention.

All orders from unknown correspondents must be accompanied with *cash or satisfactory reference.*

Hyacinths, Tulips, Tuberoses, and other Bulbous Roots imported fresh every year.

☞ *Liberal discount on large orders and to the trade.*

E. SHEPPARD & SONS,

Nurserymen and Florists,

P. O. Box 335.
Telephone.

224 FAIRMOUNT ST., LOWELL, MASS.

CATALOGUE

OF

E. SHEPPARD & SONS,

WHOLESALE AND RETAIL DEALERS IN

PLANTS,

FOR

Green-houses, Hot-houses, and Bedding.

———•———

GREEN-HOUSES AND NURSERY:

No. 224 FAIRMOUNT STREET,

LOWELL, MASS.

LARGE ENGLISH PANSY.

New Plants for 1885.

ASPARAGUS PLUMOSA.

An exceedingly handsome ornamental plant, for the greenhouse or conservatory. It has a pretty feathery growth, and is extremely useful for cutting for decorations.

Price, $2.50 each.

ASPARAGUS TENNISSIMA.

A beautiful climbing variety, of rapid growth; one of the most useful vines for decorations yet introduced, having a beautiful, light, and feathery appearance.

Price, 35 cts. to $2.00 each.

ASPARAGUS REMOSA.

A dwarf and very pretty variety, of strong growth; very useful for bouquets, cut-flowers, or floral designs of any description.

Price, 50 cts. each.

NEW JAPANESE CHRYSANTHEMUMS.

Bois Rose. Pearly white, passing to light rose; long petals, fine, large flower.

Bras-rogue. Rich, velvety, crimson-maroon; small and free.

Bend D'Or. Much twisted, grand show-flower; bright sulphur-yellow.

Bicolor. Enormous, large, flat flowers; red, striped with orange; last a long time in flower.

Ceres. Beautiful blush-pink; fine flower.

Chinaman. Brilliant violet-purple, occasionally streaked with silvery-white; fine flower.

Delicatum. Blush; very large; petals broad and flat, tapering to a point.

F. L. Harris. Bright cinnamon-red; a new and fine color; distinct and good; late.

Frinbriatum. Most delicate pink; fringed.

Gloriosum. Beautiful lemon-color; very free-flowering; one of the earliest.

Hon. John Walsh. Dark lake; a new color in Chrysanthemums; free.

H. Waterer. Enormous flowers, of great substance; flat; yellow, with copper centre.

Hiver Fleuri. Cream-color, tinted with rosy violet; centre of flower buff-yellow.

J. Collins. Immense, large, flat flowers, of copper-bronze ; a self-colored variety.
Jessica. White ; very long petals ; a great bloomer.
Lord Byron. A magnificent, large variety; very distinct orange, tipped with red.
Marquis of Lorne. Reddish carmine, spotted and tipped; reverse nankeen-yellow ; a large, bold flower, reflexed.
Mrs. Wm. Mencke. Brightest yellow, with slender petals of peculiar shape; very distinct and pleasing ; late.
Minnie Miller. Dark rose ; very free-flowering ; the best rose-colored variety.
Margot. Rosy violet, cream centre ; reverse of petals yellow ; very large and double.
M. Desbreaux. Deep chestnut-red ; extra large.
Petit Norbert. Carmine-rose, shaded silver ; the florets bordered with pure nankeen-yellow; reverse golden yellow.
Pres. Arthur. Immense rose flowers, opening in whorls; this was exhibited, measuring more than seven inches across.
Pres. Garfield. Brightest, large flowers; very distinct.
Rex Rubrorum. Deep red, shaded crimson ; reverse of petals pale nankeen.
Syringa. Lilac; immense size ; centre petals increasing, others very irregular.
Snow-storm. Pure white, distinct and free.

Price, 25 cts. each; $2.50 per doz.

NEW LARGE CHRYSANTHEMUMS.

(Incurve Varieties.)

Bruce Findlay. A model flower ; pale canary-yellow ; most distinct in color; a gem for exhibition.
David Allen. Very large ; chrome-yellow outside ; centre cinnamon-red.
Duchess. Enormous red flowers ; very free and distinct.
Golden Prince. Primrose-yellow; very free.
Gorgeous. Golden yellow; a beautiful variety ; early and distinct.
J. Lovering. Immense flowers ; white petals outside ; striped and mottled with pink inside.
Lord Alcester. Primrose ; a distinct and noble flower.
Mrs. Gladstone. Dark chestnut-red ; very fine.
Mrs. C. W. Wheeler. Immense, perfect-shaped flowers ; outside petals orange, and deep red centre.
Mrs. Geo. W. Childs. Like preceding variety in shape ; outside petals white ; dark rose centre.
Moonlight. Pure white ; flowers immense size.
Pres. Sanderson. Purple-tinted buff; flowers large and beautifully incurved.
St. Patrick. Bronzy red; fine.
W. K. Harris. Flowers ball-shaped; nankeen-yellow; at first it shows a light red centre.

Price, 25 cts. each ; $2.50 per doz.

NEW CHRYSANTHEMUMS.

(Not Classified.)

Alba Striata. Beautifully striped.
Crese Newry.

Mme. Boucharlat. Orange-brick red; reverse of petal silvery-white; a beautiful peony-formed flower.
Dormillion. Deep amaranthus; reverse of petals silvery-white; fine.
Gaillardia. Deep reddish maroon, tipped golden; large and fine.
James Townsend.
Margarette Polleriville.

Price, 25 cts. each; $2.50 per doz.

NEW SINGLE CHRYSANTHEMUMS.

Attraction. Quilled petals, of medium length; color soft, rosy lilac.
Miss Beckwith. White, shaded lilac-blush, broadly tipped with deep rosy lilac.
Miss Cannell. One of the best; finely formed; pure white; deep yellow centre; free.
Mrs. Langtry. A decided acquisition; flowers a pleasing shade, of silvery blush.
Mr. Toole. Small, well-formed flowers; clear yellow, very free, and distinct.

Price, 25 cts. each; $2.50 per doz.

SUMMER-FLOWERING CHRYSANTHEMUMS.

Sœur Melanie. Pure white; a most abundant bloomer.
Virginia. Pure white, well formed.

Price, 25 cts. each; $2.50 per doz.

NEW POMPONE CHRYSANTHEMUMS.

Brunette. Amber-yellow, shaded with reddish brown.
Fremy. A beautiful flower, with laciniated florets of a rich orange, tipped with gold.

Price, 25 cts. each; $2.50 per doz.

NEW FANCY DAHLIAS.

John Forbes. Fawn-color, striped with maroon.
Mme. Soubeyre. Beautiful rosy lilac, striped with carmine.
Polly Sandell. Beautiful lemon-color, tipped with white.

Price, 30 cts. each; $3.00 per doz.

NEW SHOW DAHLIAS.

Hope. Light rosy lilac; large and very constant.
Imperial. Deep purple, with a pretty shading of lilac; large and splendid form.
Miss Cannell. Creamy ground, deeply edged with purplish crimson.
Mrs. John Laing. French white; excellent form.
Revival. Rich crimson, full size.
Shirley Hibberd. Dark, shaded crimson, very constant.
Sunbeam. Clear buff, with a beautiful outline.
Senator. A very fine purple.

Seraph. Blush, centre well up; strong, erect habit.
Sir Garnet Wolseley. Reddish chocolate; very large and good form.
Wm. Rawlings. Rich crimson-purple, perfect outline.

Price, 35 cts. each; $3.50 per doz.

NEW SINGLE DAHLIAS.

Acquisition. Crimson, with scarlet bars at the edge of each petal; very striking.
Aurata. Very fine yellow.
Defiance. Extra fine scarlet.
Evening Star. Rich maroon.
Fire Fly. Very rich crimson.
Flavius. Good dwarf yellow.
Grandee. Large, rich purple.
Lucifer. Fine scarlet.
Purple Prince. Grand, free bloomer.
Red Gauntlet. Deep crimson-scarlet.
Striata. Lilac, striped with crimson.
Terra Cotta. Color terra cotta.
Vesuvius. Fine dwarf; scarlet.

Price, 30 cts. each; $3.00 per doz.

NEW GERANIUMS.

Advance. A splendid, bright scarlet; large, lovely shape; very smooth and round.
Ajax.
Belle Nanceined.
Boule Noire. This is the largest, darkest, and the roundest truss of any variety yet raised; the nearest to a black or purple.
Danae.

Hollyhock. This variety resembles a Hollyhock very much. Its extended front guard, or outer rounded petal, stands out beyond the double-rosette centre petals; brilliant scarlet, shaded yellow: a double guinea.

Joyful Improved. Individual flowers and trusses; very large and fine, and is what its name implies; soft magenta, with a distinct edging of orange-scarlet.

Lady Reed. A grand improvement in oculated varieties; pure white, with a large, scarlet centre; shape and size of petals almost equal to the scarlet section.

La Cygne.

Lord Derby. So large and bright, both pip and truss, that at first sight it looks like one huge mass of fiery pink.

Mrs. Bowen. A beautiful blush-color, distinct and pleasing; dwarf, bushy habit.

Mrs. Robertson. Very free and dwarf, flowers and trusses large; bright rose-pink.

Mr. J. Barriff. A most useful variety for growing into specimens, quite distinct in color; deep rosy salmon, with a very intense orange-scarlet centre; trusses large and freely produced.

Pharos. Another step towards a perfect double white, although no whiter than others; it has a larger pip and truss, with a free, fine habit.

Queen of the Belgians. This variety is certainly the best single white yet produced, and brings the whites nearly up to the size and shape of the colored varieties.

Price, 50 cts. each.

NEW BOURBON ROSE.

Malmaison Rouge. A sprout from the well-known Souvenir de la Malmaison, possessing all the qualities of that variety, with deep velvety red flowers; a most valuable variety.

Price, 50 cts. each; $5.00 per doz.

DOUBLE WHITE VIOLET.

Swanley White. By far the best Double White Violet in cultivation. It very much resembles the good old "Neapolitan" in growth, with a larger flower and flower foot-stalks, and is for choice bouquets and cut-flowers quite a gem.

Price, 35 cts. each.

General Collection.

ABUTILONS.

These plants have extremely beautiful flowers, being richly veined and striped, suitable for the garden or greenhouse; flowers bell-shaped.

Auguste Pasewold. A fine variety, foliage beautifully marked with large patches of green, yellow, and white; very distinct.

Boule de Neige. Pure white, a most abundant bloomer; flowering freely during the winter.

Darwini. Orange-scarlet, veined with pink.

King of Yellows. Clear yellow.

Mesopotamicum. Calyx scarlet, yellow petals.

"   **Variegata.** Leaves variegated yellow and green.

Purpureum. A low-growing variety, with rosy purple flowers, blooming in clusters.

Rosæflorum. Rich rose-color, veined with crimson.

Verschaffeltii. Flowers bright yellow.

Patersonii. Crimson, veined with maroon, of large size and great substance.

Sloweana Marmorata. The leaves are very large, and beautifully mottled green and yellow, the latter predominating.

Price, 15 to 25 cts. each; $1.50 to $2.50 per doz.

ACALYPHA TRICOLOR.

A very handsome plant from the Feejee Islands, with large foliage, irregularly mottled with crimson and flame-red. A beautiful plant for summer decoration. Price, 25 to 50 cts. each.

A. Marginata. 50 cts. each.

A. Macafeana. 50 cts. each.

ACHYRANTHUS.

Fine plants for vases and bedding out; particularly recommended for ribbon-gardening, etc.

Aureus Reticulatus. Leaves light green, marked with net-work of golden yellow.

Brilliantissima. A most brilliant variety, with heart-shaped leaves of a brilliant carmine, mottled with crimson, the stem and leaf-stalks pinkish violet.

Lindenii. Fine dwarf habit; lance-shaped leaves, of a deep blood-red color.
    "   **Aurea.** Leaves bright yellow, veined carmine; stems violet-crimson.
    "   **Brilliantissima.** Leaves and habit the same as A. Lindenii, but with deep carmine leaves, and deep pinkish violet stems.
Verschaffeltii. Deep crimson leaves, with bright carmine veins.

Price, 10 cts. each; $1.00 per doz.

AMPELOPSIS VEITCHII.

A small variety of the Virginia Creeper, a plant of rapid growth, and adheres very firmly to any surface. It is perfectly hardy, and like the old variety, the leaves change to a bright scarlet in the fall. Price, 25 cts. each.

    Acacia. In variety; 50 cts. each.
    Acorus Gramineus. Foliage variegated; 25 cts. each.
    Aloe Variegata. 50 cts. each.
    Ardisia Crenulata. Red berries; 50 cts. each.
    Aspidistra Lurida Variegata.

ALTERNANTHERAS.

Beautiful, dwarf-growing plants from Brazil; finely variegated with all the shades of crimson, scarlet, yellow, and green; suitable for baskets and ribbon-lines.

Paronychioides. Leaves tinted green, crimson, and straw-color.
    "   **Major.** Leaves brilliant carmine and green.
    "   "   **Aurea.** Bright golden yellow.
    "   "   "   **Nana.** Golden yellow, very compact.
Spathulata. Leaves tinted carmine and green.
Versicolor. Leaves tinted light rose to deep crimson.

Price, 10 cts. each; $1.00 per doz.

AKEBIA QUINATA.

A beautiful, *hardy* climber, attaining a height of twenty feet. Flowers chocolate-purple color; very fragrant. Price, 25 cts. each.

ARALIA FILICIFOLIA.

An ornamental stone-plant, graceful in habit, and well furnished with foliage. One of the most valuable decorative plants of its family. The stem and leaf-stocks are purplish, thickly marked with oblong white spots, with leaves expanding into a broad, leafy limb, which is imparipinnately divided. Price, 75 cts. each.

ARALIA SIEBOLDII.

A beautiful, ornamental plant, alike useful in the conservatory or the sub-tropical garden; leaves palmate, deeply lobed, bright green. Price, 50 cts. to $1.00 each.

AGERATUM MEXICANUM.

An easily-grown garden-favorite; good bedder; flowers blue.   Price, 10 cts. each; $1.00 per doz.

ALOYSIA CITRIODORA.

Well known as the "Lemon-Scented Verbena." Indispensable in every garden, for the delightful fragrance of its leaves.   Price, 15 cts. each; $1.50 per doz.

ASTILBE JAPONICA.
(Spiræa Japonica.)

One of the most beautiful of all hardy herbaceous plants, growing about eighteen inches high, with foliage of a dark, glossy green, flowering in spikes of pure white, feather-like flowers; a valuable plant for winter-flowering.   Price, 25 cts. each; $2.50 per doz.

ANEMONE HONORINE JOBERT.

A beautiful herbaceous plant, producing large white flowers from August to October; grows two to three feet high.   Price, 25 cts. each; $2.50 per doz.

ANTHERICUM VITATUM VARIEGATUM.

A useful and beautiful variegated-leaf plant; one of the best for hanging-baskets. Price, 25 cts. each; $2.50 per doz.

AGAVE AMERICANA VARIEGATA.
(Century Plant.)

No plants are more decorative or more effective than these; they require but little attention, and may be wintered under the stage of the greenhouse, or in a dry, warm cellar.   Price, 50 cts. to $1.00 each.

AZALEA INDICA.

We have a large variety of this valuable species, too numerous to describe, consisting of all the leading varieties.   Price, 50 cts. to $3.00 each.

HARDY AZALEAS.

Of all the hardy flowering-shrubs, none, perhaps, afford such a variety in color as the Azaleas, for almost every shade of pink, white, yellow, orange, and scarlet is to be found among them; and as they generally flower in great profusion, and are, many of them, deliciously scented, they deserve to be universally planted. The following varieties are perfectly hardy, and will flourish wherever Rhododendrons are grown.

Amœna. White and pink, fine.
Bessie Holdaway. Pink, fine.
Cardinal. Bright pink.
Cardoniana. Pink, light shade.
Chas. Baumann. Fine, pink.
Calendulacea Elegans. Pink, orange, and white.
Cuprea. White, shaded pink.
Fama. Rosy pink.
Fulgens. Salmon-pink.
Gloria Mundi. Bright red and orange.
Jules Cæsar. Orange and pink.
Macranth. Light yellow.
Mme. Baumann. Bright pink, striped with white and orange.

Melanie. Light rose, fine.
Meteor. Orange and brick.
Minerva. Pink and orange.
Nancy Waterer. Bright yellow, fine.
Optima. Bright orange.
Pallas. Crimson and orange.
Princeps. Fine yellow.
Pulchella Roseola. Pink.
Radiata. Pink and orange.
Roi des Belges. Rosy pink, striped with white.
Sinensis Rosea. Yellow.
Triumphans. Crimson-pink.
Unique. Fine orange.
Viscocephala. Pure white, extra.

Price, $1.50 to $3.00 each.

BOUVARDIA ALFRED NEUNER.

A new double white variety. This charming novelty will prove of inestimable value for all kinds of decorative purposes. It is of excellent habit, and a profuse bloomer, throwing large trusses of lovely, pure white, rosette-like flowers, each flower composed of three perfect rows of petals. Where cut-flowers are required, this beautiful plant is unequalled. Price, 25 cts. each; $2.50 per doz.

BOUVARDIAS.

Splendid winter-flowering and bedding plants. The following varieties are the most desirable: —

Elegans. Bright carmine, large truss.

Davidsonii. White, fine form.
Lieantha. Dark, dazzling scarlet.

Price, 25 cts. each; $2.50 to $5.00 per doz.

BEGONIA GLAUCOPHYLLA SCANDENS.

A drooping or creeping variety, with large clusters of rich salmon-colored flowers; very desirable for hanging-baskets. Price, 25 cts. each.

BEGONIA.

We have a large collection of this popular plant. The class of which B. Rex is a type, is extensively used for baskets, vases, ferneries, etc. Of the winter-flowering varieties, B. Fuchsioides, B. Weltoniensis, are representatives, producing a profusion of pink, white, and carmine flowers during the entire winter months. Price, 25 cts. each; $2.50 to $4.50 per doz.

BEGONIA.

(Tuberous Rooted.)

We have a fine collection, consisting of some of the best named varieties.   In June and July we shall have a large lot of seedlings saved from the best named sorts.   Price, named varieties, 50 cts. to $1.00; seedlings, 35 cts. each, $3.50 per doz.

BEGONIA DIAMANT.

A beautiful variety of the Rex type, with medium-sized foliage, of a rich silvery hue, tinted with rosy pink; dwarf habit.   Price, 25 cts. each.

BEGONIA METALLICA.

Perhaps of all the ornamental-foliaged Begonias, none are so beautiful as this; and for a window or drawing-room plant, it is without a rival.   Price, 25 cts. each.

BEGONIA RUBRA.

One of the finest winter-flower variety.   The flowers are scarlet-rose, glossy, and wax-like, very large and freely produced.   Price, 25 cts. each; $2.50 per doz.

BOUGAINVILLEA SPECTABILIS.

A handsome, free-flowering greenhouse climber, of very rapid growth; flowers rich mauve color.   Price, 50 cts. each.

BOUGMANSIA SUAVEOLENS.

A very showy and beautiful plant, growing from three to six feet high, with large, drooping, trumpet-shaped flowers ten inches long, white and very fragrant; blooming profusely all summer.   Price, 50 cts. each.

CALADIUMS.

A fine collection of these beautiful, ornamental-foliaged plants for the decoration of the conservatory or greenhouse in summer, with large-sized leaves assuming almost every imaginable color in their variegation.   They are of easy culture. Many of the varieties are well adapted for baskets, vases, etc.   Price, 50 cts. to $1.00 each.

Adolphe Adam. Densely speckled with white and red.
Argyrites. Spotted profusely with white.
Bicolor. Dark crimson mid-rib.
Chantini. Profusely spotted carmine, crimson mid-rib.
Devosianum. Dotted and flecked with pure white.
Duchartre. White, suffused with rose.
Houlettii. Light green, white spots, pink mid-rib.
Mme. Houlettii. Pink and white spots.
Mars. Large; crimson centre.

Pictum. Light green ground, large white spots.
Reine Victoria. White; green and crimson spots.

Price, 50 cts. to $1.00 each ; $4.50 to $10.00 per doz.

CANNAS.

These plants, by their broad, massive foliage, impart a beautiful aspect to gardens; also well adapted for pot-culture. Price, 15 to 25 cts. each ; $1.50 to $2.50 per doz.

CALCEOLARIA. --- SHRUBBY.

Bijou. Brilliant crimson, dwarf.
Golden Gem. Deep yellow, profuse bloomer.

Price, 10 cts. each ; $1.00 per doz.

COLOCASIA ESCULENTA.

One of the most attractive ornamental-foliaged plants in cultivation. The leaves are heart-shaped, of immense size. As a single plant for the lawn or large flower-borders, it has no superior. Price, 25 to 50 cts.; $2.50 to $5.00 per doz.

MONTHLY CARNATIONS.

Adonis. Variegated, red and white.
Astoria. Yellow, scarlet, and white.
Fred Johnson. Brilliant scarlet.
Garfield. Bright crimson-scarlet.
Grace Wilder. A beautiful clear pink, finely fringed.
Gen. Grant. Pure white, in clusters.
Henrietta. Striped rose and purple.
La Purite. Rosy carmine, profuse bloomer.

Miss Joliffe. A delicate blush-pink.
Princess Louise. Bright carmine-rose ; quite fragrant ; free bloomer.
Pres. Degraw. Pure white, fragrant.
Snowdon. Snow-white.
Smith's Seedling. Large, white, fragrant.
Variegated La Purite. Carmine, striped.

Price, 10 cts. to 25 cts. each ; $1.00 to $2.50 per doz.

CENTAUREA CANDIDISSIMA.

A valuable plant for massing, or to contrast with dark-colored foliage in ribbon-rows; leaves downy white; compact habit. Price, 15 cts. each ; $1.50 per doz.

CENTAUREA GYMNOCARPA.

A beautiful plant, with silvery gray foliage; well adapted to contrast, in ribbon-lines, with dark-foliaged Coleus or Achyranthus. As a basket-plant it is unsurpassed, its drooping, fern-like leaves being very effective. Price, 10 cts. each ; $1.00 per doz.

CERASTIUM TOMENTOSUM.

A white-foliaged plant of trailing habit; well suited for hanging-baskets or stands. Price, 10 cts. each; $1.00 per doz.

COROZEMA VARIA.

A beautiful greenhouse shrub, bearing purple and orange-colored flowers in spikes four to six inches in length, lasting in bloom through January and February. Price, $1.25 each.

CHRYSANTHEMUMS.

(Large Flowering.)

Alfred Salter. Delicate silvery pink; large and fine.
Blonde Beauty.
Bronze Jardin des Plantes. Bronze-orange; yellow centre.
Duke of Roxburgh. Bright yellow.
Duchess of Manchester. White, large flower.
Empress of India. Clear white.
Ellen Turner. Lilac-rose.
Gen. Halley. Light rose.
Genice. Rosy pink.
Guernsey Nugget. Primrose-yellow.
Inner Temple. Magenta-crimson, distinct.
Jardin des Plantes. Golden yellow.
Julia Lagravere. Velvety crimson; fine.
Lord Wolseley. Sprout from Prince Alfred, of more substance, with bronzy hue.
Lord Stanley. Buff.
La Purite. Creamy white.
M. Lucian Balk. Reddish crimson and orange.
Mlle. Corizette. Deep rose, shading off to a beautiful silvery rose.
Mabel Ward. Canary yellow; very fine.
Miss Mary Morgan. Delicate pink, fine.
Mrs. Geo. Rundel. Pure white, incurved flower.
Mrs. Parnell.
Mrs. Dixon. Splendid yellow.
Mr. Jay. Reddish crimson, fine.
Ne Plus Ultra. Rose.
Pearl des Beauties. Deep crimson and brilliant amaranth.
Prince Alfred. Rose-crimson, large and fine.
Prince of Wales. Dark purple-violet.
Queen Victoria. White.
Rose Perfection. Bright rose.
Sunflower. Pale canary-color.
Snowball. White, ball-shape.
Tragedie. Rosy violet; flowers full.
Venus. Delicate peach.
Volunteer. Rosy lilac.
White Venus. White, finely incurved.

Price, 15 cts. to 25 cts. each; $1.50 to $2.50 per doz.

JAPANESE CHRYSANTHEMUMS.

Agrements de la Nature. Golden yellow, shaded reddish brown; very fine.

Bouquet Fait. Rich rose and silvery white, centre shaded yellow, flowers large and very fine.

Comtesse de Bomgard.

Chang. Yellow, tipped with crimson.

Comedie. Beautiful clear rose, medium size, and very double.

Comtesse de Beauregard. Light rose; large flower.

Dr. Masters. Centre bright red, tipped with gold.

Elaine. Pure white, very full; fine.

Fee Rageuse. White, shaded with lavender; large.

Fair Maid of Guernsey. Pure white, extra large.

Grandiflora. Large, golden yellow; very fine.

Henry Brock. Purple-crimson.

Helen McGregor. Orange and crimson, small.

James Salter. Pink, changing to white.

Jonas.

Lady Selborne. Pure white; very fine and useful.

L'Infante D'Espagne. An immense flower; beautiful pale yellow.

L'Isle des Palisirs. Brownish crimson, shaded orange and gold; free.

Lord Beaconsfield. Salmon-red, shaded with amber yellow, reverse of petals nankeen yellow; remarkable and distinct.

Meg Merrilies. Sulphur-white, very large; most curious form; late.

Mary Major. Beautiful creamy white; a large, grand flower.

Mrs. Chas. Cary. White, much curled, large flowers; one of the finest.

M. Baco. Vivid crimson and orange, with golden centre.

M. Ardine.

M. Richard Larois. Dark rose and bright violet, tipped pure white; an enormous flower.

Orphee. Brick-red and crimson, centre gold, incurved flower.

Purple King. Deep purple.

Reine des Beauties. Deep rich crimson, reverse of florets golden; free.

Reve de Printemps. Rich dark violet-crimson, reverse of florets silvery white; large.

Striatum Perfectum. A fine, large flower, white; beautifully striped and flaked violet-rose.

Souvenir de la Marrides.

Triomphe du Chatelet. Salmon-color, tinted rose, golden centre, immense size.

Ville de Hyeres. Yellow, striped brown; fine.

Price, 10 cts. to 25 cts. each; $1.00 to $2.50 per doz.

POMPONE CHRYSANTHEMUMS.

Admiral.

Apotheose. Purple-carmine.

Arbre de Noel. Bright orange, tinted red; imbricated.

Anna du Belocca. Sulphur-white.

Amphillia.

Damiette. Blush, dark centre.

Duruflet. Rose-carmine.

Gen. Canrobert. Pure yellow.

Golden Circle. Golden yellow, quilled flower.

La Fiancee. White, fringed.

La Mascotte. Mahogany-color; imbri-
cated.
Model of Perfection. Rich lilac-
edge, pure white, distinct.
Mme. Domage. White, early, fine.
Maroon Model. Maroon-crimson.

Mlle. Martha. Pure white, extra fine.
Nellie. Creamy white.
Rose Mantle. White, pink-shaded.
Volcan. Flowers small; reddish ma-
hogany.

Price, 10 cts. to 25 cts. each; $1.00 to $2.50 per doz.

CHRYSANTHEMUMS.

Very early-flowering varieties.

Early Cassey. Light, tipped lilac.
Indicum Nanum. Creamy white.
Mme. Dufoy. Plain white.

Mrs. Campbell.
Precocite. Bright yellow.
Scarlet Gem. Maroon-red.

Price, 10 cts. to 25 cts. each; $1.00 to $2.50 per doz.

CISSUS DISCOLOR.

A handsome hot-house climber, with leaves beautifully shaded with green, purple, and white. It should be grown in a moist, warm atmosphere, and partially shaded from direct rays of the sun. Price, 25 cts. each; $2.50 per doz.

CINERARIA MARITIMA CANDIDISSIMA.

A beautiful silvery-foliaged plant, compact habit. Price, 10 cts. each; $1.00 per doz.

COBŒA SCANDENS.

A well-known climbing plant of vigorous growth, and may easily be trained to a height of thirty feet, if desired. Price, 25 cts. each; $2.50 per doz.

COBŒA SCANDENS VARIEGATA.

A beautiful variety of the above. The leaves are margined with yellowish white, which form a beautiful contrast to its large, purple, bell-shaped flowers. Price, 25 cts. each.

COPROSMA BAUERIANA VARIEGATA.

An exceedingly handsome plant, with bright, glossy, green-and-yellow leaves; excellent for decorating; also a good bedder. Price, 35 cts. each.

COLEUS, GOLDEN BEAUTY.
(Golden Bedder.)

The most useful Coleus introduced since C. Verschaffeltii. Its beautiful golden yellow leaves, very dwarf branching habit, and its remarkable color in hottest sun, make it indispensable where the best bedding is required. Price, 10 cts. each; $1.00 per doz.

COLEUS.

(General Collection.)

The Coleus are now too well known to require any description. Valuable for massing or ribbon-borders, in contrast with lighter foliage. We have selected the following as the most distinct of the named varieties.

Beacon. Blackish-purple, with brown and bright crimson mid-ribs and veins; foliage large.

Bismarck. Rich, velvety, dark crimson centre, golden yellow edge; dwarf.

Dr. Joe Hooker. Dark crimson, stained with dark brown; narrow, dark green margin.

Eclipse. Scarlet, shaded with brown; yellowish green, serrated edge.

Golden Circle. Rich, bronze-crimson centre; broad yellow margin; one of the best.

Golden Gem. Crimson-bronze, margined with bright sulphur-yellow.

James Barnshaw.

Kentish Fire. Centre deep crimson, bronze zone, green edge.

Lord Falmouth. Centre pale yellow, zoned with carmine, yellowish-green edge.

Miss Retta Kirkpatrick. Large white centre, shaded with yellow; broad, green, lobed margin; large foliage.

Miss Rosina.

M. J. Linden. Bronzy crimson centre, broad, bright yellow edge.

Nellie Grant. Light crimson centre, deep yellow edge, distinct.

Paroquet. Yellow, maculated with crimson and green.

Red Cloud. Rich crimson, evenly marmorated with blackish brown, narrow green margin.

Seraph. Fiery crimson, spotted with chocolate, bright green edge.

Superbissima. Blackish maroon, with a brilliant, broad purple band through the centre of the leaf.

Verschaffeltii. Rich dark crimson, finest bedding sort.

Zephyr. Rich, bronze crimson, slightly marbled with dark olive-green, violet-purple veins.

Price, 10 cts. to 15 cts. each; $1.00 to $1.50 per doz.

CLEMATIS INDIVISA LOBATA.

A beautiful greenhouse variety, producing an abundance of white flowers; very useful. Price, 50 cts. each.

CLEMATIS.

(Virgin Bower.)

The Clematis are beautiful, free-growing, fast-climbing, and perfectly hardy vines. They are well adapted for training on trellis-work, and grow from ten to fifteen feet high.

Albert Victor. Deep lavender, with brown rib, changing to white.

Mrs. S. C. Baker. French gray, one of the best.

Purpurea Elegans.

Rubella. Beautiful, rich claret-purple.

Sir Garnet Wolseley.

3

Beauty of Surrey.
Countess of Lovelace.
Fair Rosamond.
Flamula. White, sweet-scented.
Gem. Deep lavender-blue.
Gipsy Queen.
Jackmanii. Violet-purple, veined centre.
John Gould Veitch. Light blue, large and double.
Lanuginosa Nivea. Pure white, fine.
Lord Londesborough. Deep mauve, purplish-red band.
Lucie Lemoine. Double, white, extra.
Miss Bateman. Fine white.
Patons Hybrid Siebaldii. Lavender-blue, very large, and continues in bloom
    a long time.

Price, 50 cts. to $1.00 each.

CROTONS.

Handsome hot-house plants, with very ornamental foliage.

Acubafolia.	**Mooreanus.**
Angustifolia.	**Multicolor.**
Aurea Maculata.	**Nobilis.**
Cornutum.	**Pictum.**
Disraeli.	**Prince of Wales.**
Elegans.	**Queen Victoria.**
Evansianus.	**Variegatum.**
Fasciatus.	**Veitchii.**
Interruptum.	**Volutum.**
Irregulare.	**Weismanii.**
Macarthurii.	**Youngii.**
Maximum.	

Price, 50 cts. to $2.00 each.

CURCULIGO RECURVATA.

A noble plant, of Indian origin. The leaves are long, stalked, spreading, lanceolate, longitudinally plated leaves, beautifully recurved, extremely useful for decorative purposes. Price, 50 cts. to $2.00 each.

CURCULIGO RECURVATA VARIEGATA.

This is a remarkably handsome and ornamental stove-plant, producing from a
tuberous rhizoma, an arching head of recurved, plated, oblong-lanceolate leaves
upwards of two feet long and six inches wide, on stalks a foot and a half in length.
The leaves are green, banded in a varying manner, with clear white stripes. Price,
$1.00 to $2.50 each.

CYCLAMEN PERSICUM.

A fine collection of these beautiful winter-blooming plants, comprising all the
shades of color from pure white to deep crimson, many of the varieties being

striped and blotched. Price for good, strong, flowering plants, 50 cts. each; smaller plants 25 cts. each; $2.50 per doz.

CYPERUS ALTERNIFOLIUS VARIEGATUS.

A beautiful grass-like plant, throwing up stems to the height of three feet, surmounted by a cluster or whorl of leaves. The stems and leaves are beautifully variegated with white; splendid plant for aquariums, fountains, baskets, etc. Price, 25 cts. to 50 cts. each.

SHOW DAHLIAS.

Our varieties of this most beautiful and showy fall flower have been selected with special reference to constancy and continuance of bloom. They are all first-class varieties. Many of the new and high-priced varieties of last season are included in this selection.

Bob Ridley. Red, good centre.

Champion Rollo. Large, dark orange, with a light shade on the edge of the petals; fine.

Christopher Schmidt. Light salmon, fine form.

Chas. Leicaster. Beautiful, bright scarlet.

Constancy. Yellow ground, deeply edged with lake; a telling flower.

Dr. Rozies. Bright scarlet, fine form.

Duke of Connaught. Dark crimson, sometimes shaded purple; very constant; good form.

Duke of Albany. Rich crimson, fine form; free.

Georgina. Large, creamy white; a well-built flower.

Geo. Dickson. Chestnut-brown; very constant.

Herbert Turner. French white, very large, and finest form; constant.

H. W. Ward. Yellow ground, heavily edged and shaded with crimson; fine form.

John Bennett. Yellow, deeply edged with scarlet.

John McPhersons. Violet-purple, fine form.

James Vick. Purple-maroon, color intense, very full.

John Stephen. Bright orange-scarlet, new in color; finest form.

Lady Wimborne. Deep pink, heavily shaded with rose; very pretty; a new color.

Laura Livingston.

Louisa Neat. Delicate pink, creamy white centre; splendid form.

Lizzie Liecaster. Pink, curiously pencilled; constant.

Miss Henshaw. Very large, white, full centre; constant.

Mrs. Sibbald. Primrose, deeply edged with lake; a desirable color.

Maggie Soul. Blush-white, edged with purple; profuse-flowering.

Miss M. Batchellor. Beautiful bright scarlet; a perfect model in shape.

Model. White, full centre.

Mrs. Hodgson. Yellow ground, heavily edged with crimson; good petal; centre and outline grand.

Mrs. Seaman. Yellow ground, edged with lake.

Nellie Buckle. Lilac-shaded rose, full and fine.

Pre-eminent. Purple or plum-color; extra.
Queen of Beauties. Pale straw, beautifully tipped with purple; fine form.
Wm. Davis. Light purple; a fine, early variety.
Wm. Keynes.
Walter Reid. Purple, with magenta tinge; fine flower, full and large.

Price, 20 cts. to 35 cts. each; $2.00 to $3.00 per doz.

POMPONE, OR BOUQUET DAHLIAS.

Ardens. Clear straw.
Dr. Schwabes.
Fannie Weiner. Yellow, with a light crimson edge.
Fred Muller. Reddish buff.
Fire-ball. Bright orange-red.
Floribunda. Rich carmine-red.   .
Gem. Crimson.
Gold Meteor. Golden yellow.
Karl Goldenburg. Small, yellow, distinctly tipped with white.
Keleine Moulettin. Maroon.
Keleiner Serius. Scarlet.
Linda. Orange.
Little Dear. Blush-white, tipped with rose.
"   **Elizabeth.** Rosy lilac, tipped.
"   **Gold-light.** Creamy white.
"   **Helen.** Light blush.
"   **Kate.** Velvety purple.
"   **Wonder.** Crimson-scarlet.
Peasant Girl. White, edged with crimson.
Prince of Lilliputians. Maroon.
Pure Love. Lilac.
Rubincentiflora. Dark maroon, almost black, scarlet tip.
Sacramento. Yellow, edged with red.
Sappho. Dark crimson.
Schmitz' Defiance. Yellow, white tip.

Price, 15 cts. to 25 cts. each; $1.50 to $2.50 per doz.

FANCY DAHLIAS.

Annie Pritchard. White, beautifully striped with lilac-rose; a large flower.
Golden Eagle. Yellow, fine, laced with purple.
Jessie McIntosh. Red, distinctly tipped with white.
John Forbes. Fawn color, striped with maroon; one of the finest.
Lady Antrobus. Red, tipped with pure white; fine form.
Prof. Faucett. Dark lilac, striped with chocolate.
Prospero. Crimson, tipped with purple.
Polly Sandell. Beautiful lemon-color, tipped with white.

Price, 25 cts. to 35 cts. each; $2.00 to $3.00 per doz.

BEDDING DAHLIAS.

Flag of Truce. White, flaked lilac.
Mt. Blanc. White, free-flowering.
Queen Victoria. Fine yellow.

Price, 25 cts. each; $2.50 per doz.

DAHLIA JAUREZII.

A most valuable and useful decorative plant for the late summer and autumn months. Its blossoms are of a rich crimson, and very much resemble, in shape and color, the well-known cactus, "Cercus Speciosimus." Height about three feet, very bushy; flowers of very striking appearance, quite unlike those of any ordinary double dahlia, the florets being flat and not cupped. Price, 25 cts. each; $2.50 per doz.

SINGLE DAHLIAS.

Amber. Amber, tinged purple; free.
Auranciaca. Orange-scarlet.
Canary. Bright canary-yellow, extra fine.
Coccinea Buffalo. Pale dun or fawn-color, flushed with orange-red.
      "   **Cato.** Clear orange-scarlet. The florets are yellow at the base.
Comet. Deep reddish scarlet, veined and suffused with orange and gold flowers; medium size, fine shape and form.

Cloth of Gold. Rich yellow.
Cecil Teigner. Rich, rosy pink; free.
Conqueror.
Dido. Magenta-rose, paler at the tips, with a yellow zone at base.
Gracilis Perfecta. Velvety crimson, free bloomer.
Halo. Magenta-crimson; a zone of yellow at base of florets.
Harlequin. Deep rose ground-color, having a broad band of purple down the centre of each petal; very attractive.
Harold. Very deep rose.
Irene. White, delicately-shaded pink.
Le Baron. Rich mulberry, shaded crimson, having a dark line around the centre.
Lovely.
Lutea Grandiflora. Rich yellow, large, well-shaped flower; very free, and good habit.
Lutea. Bright yellow.
Mrs. Cullingford.
Maud Willan. Delicate straw, shading to creamy white.
Nora. Bright pink, a very free and effective bedder; extra.
Paragon. Dark maroon.
Purple Paragon. Violet-purple; habit, etc., same as Paragon.
Ruby. Rich ruby-red, fine.
Rob Roy. Intense, bright scarlet, very early; extra.
Scarlet Cervantesii. Bright scarlet.
     " Gem. Bright scarlet.
Single Zinnia. Rich crimson-scarlet; a neat flower, dwarf habit.
Solfaterre. Pale primrose.
Thalia. Pretty, rich amaranth, very dwarf; extra fine.
Thos. Wheeler. Pretty terra-cotta color; a good-shaped flower.
Tyro. Lilac-yellow zone at the base.
Wm. Gordon. Mauve, flushed with pink.
White Queen. One of the best and most useful.
Zulu. Deep maroon, shaded crimson; color intense.

Price, 15 cts. to 25 cts. each; $1.50 to $2.50 per doz.

DAISIES.

A fine collection of these handsome, spring-flowering plants, of various colors. Price, 10 cts. each; $1.00 per doz.

DIPLADENIAS.

Brearleyana. An evergreen stove-climber of the finest rank; its oblong, dark green leaves serve as a fine contrast to the beautiful flowers, which are of the largest size, opening pink, changing to the richest crimson-color.
Boliviensis. This beautiful variety is one of the most useful. The flowers are medium size, pure white, with a distinct yellow centre. Price, 50 cts. to $1.00 each.

DRACÆNAS.

Beautiful ornamental-leaved plants, much used as centre plants in rustic baskets, vases, etc., and are among the most beautiful plants for table decoration.

Amabilis. Ground-color bright, shining green, suffused with creamy white and rosy pink.
Baptistii. Light-green foliage, streaked with bright rose.
Congesta. Narrow, green leaves, rapid growth.
Cooperii. Large, purple-and-red foliage.
Ferrea. Broad, dark-brown leaves.
Gracilis. Green foliage, of a graceful habit, with drooping leaves.
Guilfoylei. Bright-green foliage, striped pink and white.
Hendersonii. Distinctly striped with white and rosy pink.
Hybrida. Deep green, margined with rose.
Imperialis. Dark ground, changing to bright red and light rosy pink.
Macleayi. Very dark bronze, with metallic lustre; dwarf.
Magnifica.
Mooreana. A bright, reddish-crimson color, of graceful habit.
Nobilis. High-colored variety, leaves greenish bronze.
Nobilis Stricta. Marked with crimson.
Porphyrophylla. Deep bronzy-red hue on the upper side, and slightly glaucous beneath.
Rubra Elegans. A narrow-leaved variety, striped red, margined with rose.
Shepperdii. Young leaves dark green, striped with paler/green, which changes with age to orange-red.
Superba. Dark bronzy red, changing to bright crimson.
Terminalis. Bronze-red, changing to bright crimson.
Veitchii. Narrow green-leaved variety, of graceful habit.
Youngii. Young leaves light green, tinged with rose, changing to copper-color.

Price, 50 cts. to $1.00 each.

DIEFFENBACHIA BAUSEI.

A stalky-growing variety, with broad leaves, of dwarf habit, color of leaves a yellowish-green hue, irregularly edged and blotched with dark green, and spotted with white. Price, 50 cts. to $1.50 each.

ECHEVERIAS.

This beautiful and interesting genus of succulent plants is now attracting unusual attention. They are desirable as pot-plants for decorative purposes.

Atropurpurea. Long, purplish leaves.
Metallica. A very ornamental plant, with large, massive leaves of a beautiful metallic hue.
Retusa Secunda.
Sanguinea Miniata Nuda.
Secunda Glauca. A small, compact-growing variety, with green leaves.

Price, 10 cts. to 50 cts. each.

ERANTHEMUM TRICOLOR.

This plant requires a warm temperature to bring out its variegations, which are pink, maroon, and purple. Valuable for massing. Price, 25 cts. each.

ERANTHEMUM ATROPURPUREUM.

A beautiful variety, with rich, rosy-purple leaves, requiring the same treatment as *E. Tricolor*. Price, 25 cts. each.

ERIANTHUS RAVENNÆ.

A tall, ornamental, reed-like grass, having a beautiful foliage and dense, silvery-white plumes. Being perfectly hardy, it will prove a most desirable plant for the decoration of lawns. Price, 25 cts. each; $2.50 per doz.

EULALIA JAPONICA VARIEGATA.

A very ornamental grass, of easy culture, perfectly hardy. The leaves are long and narrow, and striped green-and-white. The flower-stem is from four to six feet high, terminating with a head or pannicle of blossoms which, when cut off, retain their beauty a long time. Price, 50 cts. to $1.00 each.

EUCHARIS AMAZONICA.

A very handsome green-house plant. The flowers, which are freely produced, are star-shaped, pure white, deliciously fragrant, four inches across. Easily grown in a warm, moist atmosphere. Price, $1.00 to $2.00 each.

EXOCHORDA GRANDIFLORA.

A beautiful, hardy shrub, growing about four feet high; producing beautiful spikes of white flowers in May and June. Price, $1.00 each.

FUCHSIAS.

Aurora. Salmon; long tubes and sepals.

Avalanche. Bright carmine sepals, violet corolla, flowers double.

Berenice. Tube and sepals rich crimson; the sepals broad and beautifully reflexed.

Brilliant. Corolla scarlet, sepals white.

Carminata. Tube and sepals carmine, the latter very broad and well reflexed, deep violet corolla; the petals are very round and smooth.

Charming. Scarlet tube and sepals, purple corolla.

Clarinda. Huge-spreading double white corolla, which is exceedingly attractive; short tube and broad sepals of a rich crimson.

Clipper. Tube and sepals carmine-scarlet and well reflexed; corolla deep purple, shaded violet.

Coccinia. Scarlet, self-color.

Creusa. Rich crimson tube and sepals, the latter short and completely reflexed; large corolla, a rich dark purple plum-color, shaded crimson at the base.

Earl of Beaconsfield. Tube and sepals rosy carmine, corolla deep crimson; free-blooming variety.

Eclipse. Large corolla of a deep purple, fine shape; bright red tube and sepals, latter broad and well reflexed; good habits, free-flowering.

Fairest of the Fair. Sepals and tubes white, rich velvet corolla.

Grand Duchess Maria. White tube and sepals, rose corolla; strong grower.

    "   "   Large, pure white, double corolla, of fine form; brilliant carmine tube and sepals.

Gustave Dore. Rich carmine tube and sepals, double white corolla; medium size and a free bloomer.

Jules Ferry. Scarlet tube and sepals, violet-mottled white corolla; a pretty variety.

Jennie de Arc. Pure white corolla, tube and sepals bright red; extra.

J. J. Rosseau.

Louis Blanc '

Lord Wolseley. A fine, large flower, with broad, reflexed sepals, vinous red color. Corolla, soft rosy crimson, veined with red and margined with bluish purple; excellent habit, and a profuse bloomer.

Mrs. Rundell. A grand improvement on the well-known Earl of Beaconsfield; very quick, strong growth, and very free-flowering; a grand market variety.

Miss Lizzie Vidler. Light rosy-red sepals, beautiful soft mauve corolla; very free.

Mrs. H. Cannell. Scarlet tube and sepals; large, double, pure white corolla; the finest double white.

Noveau Mastidonte. Excellent habit, and of free growth; flowers full and double; the corolla globular, color dark violet, veined with red; sepals beautifully reflexed.

Princess Beatrice. The tube is rather stout and short, sepals deep crimson, corolla reddish purple, good-shaped flower.

President. Tube and sepals bright vermilion; corolla a very rich violet.

Rose of Denmark. Tube and sepals blush, the latter beautifully recurved; corolla delicate pink.

Rose of Castile. Blush-white sepals, corolla rosy-purple.

Speciosa. Fine, early-market variety.

Vainqueur de Puebla. Sepals bright red, corolla white.

Venus de Medici. Tube white, sepals blush-white, corolla blue.

Wave of Life.

Price, 25 cts. each; $2.50 per doz.

FERNS.

No plants are more general favorites than ferns. Their great diversity and gracefulness of foliage make them much valued as plants for ferneries, baskets, rock-work, etc.

Adiantum Affine.	**Adiantum Farleyense.**
" **Amabile.**	" **Formosum.**
Cuneatum.	" **Macrophyllum.**

4

Anemia Flexuosa.
Asplenum Longissimum.
Cheilanthes Elegans.
Davallia Canariensis.
   "   Mooreana.
   "   Tenuifolia.
Doodia Caudata.
Gymnogramma Chrysophylla.
   "   Peruviana Argyrophylla.
Gymnogramma Tartarea.
Lomaria Gibba.
Lygodium Scandens.
Microlepia Strigosa.
   "   Herta Cristata.
Nephrodium Mollie Corymbiferum.
Nephrolepis Exalata.
   "   Davallioides Furcans.
Phlebodium Aureum.
Platycerium Alcicorne.
Platyloma Rotundifolia.

Polypodium Vacinifolium.
Polystichum Proliferum.
Pteris Argyrea.
   "   Cretica.
   "   "   Alba-Lineata.
   "   Hastata.
   "   Longifolia.
   "   Scaberula.
   "   Tremula.
   "   Tricolor.
Selaginella Circinalis.
   "   Cæsium-Arborea.
   "   Densa.
   "   Flabellata.
   "   Japonica.
   "   Kraussiana.
   "   Martensii.
   "   Poulterii.
   "   Robusta.
   "   Rubricaulis.
   "   Umbrosa.
   "   Wildenovii.

Price, 15 cts. to 50 cts. each; $1.50 to $5.00 per doz.

TREE FERNS.

Alsophila Australis.   Cyathea Dealbata.

FICUS ELASTICA.

A very ornamental plant, with very large, thick, glossy-green leaves; a fine decoration plant for the garden in summer, or the conservatory in winter. Price, $1.00 to $3.00 each.

GARDENIA FLORIDA.

(Cape Jessamine.)

Plants with dark, glossy foliage; flowers pure white, blooming in spring, deliciously fragrant. Price, 50 cts. each.

GESNERA EXONIENSIS.

A valuable winter plant of great beauty. Flowers intense orange, in dense spikes; leaves a rich, velvety purple, studded with minute red hairs. Price, 50 cts. to $1.00 each.

GNAPHALIUMS.

Neat, white-leaved plants, suitable for baskets, vases, or narrow ribbon-lines.

G. Lanatum. Downy-white foliage, of a vigorous growth.
G.   "   Variegatum. Same style of growth, with variegated leaves.

G. Saundersonii. Very shrubby, growing about six inches high. Silvery white.
G. Tomentosum. Narrow, lanceolate leaves, forming a bush ten to twelve inches
high.

Price, 10 cts. each; $1.00 per doz.

GLOXINIAS.

Of these magnificent summer-blooming plants, we have a splendid collection,
both of upright and pendant varieties. They are well worthy a place in every
collection. Price, 50 cts. each; $3.00 to $4.00 per doz.

GLADIOLUS.

GLADIOLUS.

No plant blooms so freely, with little or no care, as the Gladiolus. The varieties
are now so numerous, and many of them so much resembling each other, that we
do not give a descriptive list of varieties; no garden, however small, should be
without them. We offer, this season, a choice collection of seedlings, surpassing
many of the named varieties in beauty, and at one-half the price. Price—Named
varieties, 25 to 50 cts. each; $3.00 per doz. Seedlings, 10 cts. each; $1.00 per doz.;
$5.00 per hundred.

CHOICE ZONALE GERANIUMS.

Clytie. Crimson, suffused magenta, edged with orange-crimson, white eye;
flowers large and fine.

Emperor. Bright crimson-scarlet, flowers large, and trusses very fine.

Edith George. Rich reddish pink, trusses very large and globular, fine overlapping petals.

Favorite. Cerise-scarlet of a most pleasing tint, flowers large and of the finest form.

Ferdinand Kauffer. Deep, rich magenta-purple, orange-scarlet in upper petals; very bright and distinct.

Imogen. Salmon, of a beautiful soft shade, flowers and trusses well formed.

Kate Farmer. Soft rosy salmon, flowers and trusses finely formed.

Kate Greenway. Lovely, large, round-shaped flowers, of a beautiful bright pink.

Mr. C. L. Teesdale. Scarlet, distinct; size of pip and truss monstrous.

P. N. Fraser. Bright scarlet, small white eye, flowers and trusses large and well formed.

Surrey Scarlet. Soft scarlet, white eye, fine form and substance.

Snowball. A pure white in all seasons, pip large and round; trusses large. It is much the finest white yet introduced. Price, 75 cts. each.

Zelia. Rich crimson, tinted purple and orange; flowers large, trusses enormous.

Price, 35 cts. each.

GERANIUMS.

(Zonale General Collection.)

Brutus. Dark scarlet; very fine, single.

Chieftain. Orange-scarlet.

Cleopatra. Flowers of a brilliant shade of carmine-scarlet, trusses of fine form; very free-flowering.

Corregio. Crimson, tinted purple; fine.

Dante. Deep rose-pink, white centre.

Diana. Orange-scarlet, large zone.

Duchess. Carmine-scarlet, large truss.

Eureka. The finest single white; trusses large, and stand well above the foliage.

Fanny Catlin. A beautiful variety. Color, soft rosy salmon; large white eye; immense trusses; a strong grower.

Gertrude. Light salmon.

Gathorn Hardy. Orange-scarlet; very fine.

Guinea. Brilliant orange-yellow, very distinct.

General Grant. Brilliant scarlet, fine bedding variety.

Glow. Orange-scarlet, dwarf habit.

Globosa Major. A deep, maroon-crimson flower, trusses of immense size and perfectly globular form.

Helen Dick. Pale salmon-pink.

Incomparable. Orange-scarlet, white eye.

International. Rich crimson or crimson-scarlet, fine.

Jealousy. A very distinct color, orange-scarlet, of a yellowish hue.

Joyful. Light magenta, distinctly edged with orange-scarlet; flowers and truss large; fine habit.

La Clain. A beautiful salmon.

Lady Kennedy. Intense glowing scarlet, white eye, flowers single and perfect shape.

Laura Strachen. Deep salmon, large trusses; free.
Louis Valliott. Dark crimson-scarlet, dwarf habit.
Jean Ill. Pink, shaded with very deep purple; immense trusses.
Lizard. Rosy salmon, with a distinct bright ring of red; very dwarf; flower single.
Mrs. Strutt. Purplish pink; flowers and trusses large and well formed; a pleasing color; dwarf and free.
Mrs. Moore. Pure white, with a beautiful, distinct ring of scarlet; small white eye.
Manfred. Light orange-red, fine form.
Madonna. Deep rosy pink, white on upper petals; large.
Master Christine. Deep pink, very large trusses.
Mme. Voucher. Large, white, best.
Miss Nelson. Rosy pink, fine bedding variety.
Mrs. Durrell. Beautiful carmine-pink, large trusses, and stand well above the foliage.
Nemesis. A beautiful nosegay, color a very dark scarlet; trusses large.
New Guinea. Orange-yellow, two shades lighter than Guinea; dwarf, compact habit; flowers well formed.
Olive Carr. Rosy pink; flowers large, and good shape.
Placci. Scarlet, shaded magenta.
Placida. Deep pink, large, white centre, single.
Rev. A. Newby. White, pink eye.
Riceii. Bright scarlet, white eye.
Silvio. Bright orange-scarlet, white eye.
Sophia Berkin. Bright mottled salmon; trusses large and freely produced.
Sunbeam. Brilliant scarlet; bright orange hue.
Spencer. Magenta, white eye; a very free-bloomer.
Salmon Rienzi. A beautiful salmon self, of good size and substance.
Titania (Pearson's). Bright salmon, distinct; fine for winter blooming.
Vesuvius. Scarlet dwarf.
Wheel of Fortune. Soft rosy scarlet; petals broad; fine-shaped flowers.
Wm. Pearsons. Fine, dark scarlet.

Price, 20 cts. to 35 cts. each; $2.00 to $3.50 per doz., except where noted.

GERANIUMS.

(Double.)

This class forms an interesting and valuable addition, not only from their novelty, but from their great value as bedding-plants. As the blossoms do not shed their petals, they remain a long time in flower, and are useful in various ways where single varieties are almost worthless.

Arethusa. Deep scarlet, fine.
Bataclan. Rich dark carmine, shaded purple.
Bishop Wood. Upper petals violet, lower orange-scarlet; abundant bloomer.
Carillon. Deep salmon, of a pleasing shade, large and fine.
Depute Brice. Large flowers; trusses large; petals the color of peach-blossoms, with tints of carmine; very novel.

Depute Berlet. Beautiful shade of pink, deep tinge of violet.

"   **Varroy.** Enormous, rather flat trusses; color vivid lake, deep at the edges of petals, and shading to white at the centre; a magnificent shade.

"   **d'Ancelon.** A beautiful, well-shaped flower, color deep magenta-rose, large truss; a good grower.

Dolable. Brilliant scarlet, large truss.

Emily Laxton. Immense, vivid scarlet, velvety, very double, almost quilled.

Gen. Ferre. Orange-salmon, flowers large and beautifully mottled; trusses very large, and produced in abundance.

Henri Cannell. Deep magenta-purple, large pips, full and double, highly suffused, purple.

Henry Burrier. Fine orange-salmon.

Heteranthe. Bright scarlet, semi-double, fine form.

Heroine. The largest pip of any in the class; pure white; very fine.

J. C. Rodbard. Salmon-red.

Louis Poitier. Deep salmon; guard petals blotch-white; novel and beautiful.

M. Pasteur. Deep crimson; flowers large and well formed.

Monsieur Gelein Louagie. Semi-double, very large ; color bright vermillion, shaded with salmon; dwarf habit.

Mrs. Mampion. Rose-pink, large flower.

Maria Bertier. White, shaded pale pink, large trusses.

Mme. Gibbard. Brilliant pink.

"   **Thibault.** Very free-flowering, trusses large and fine, flowers very large and double, bright magenta-rose, upper petals marked with white.

Souvenir de Carpeux. Deep magenta; good habit.

"   " **R. Mason.** White.

Vesta. Bright scarlet, small trusses, free-flowering.

Price, 20 cts. to 35 cts. each; $2.00 to $3.50 per doz.

NEW VARIEGATED-LEAVED GERANIUMS.

Mme. Salleroi. A distinct variety, with leaves from one to two inches in diameter; the centre of each is of deep olive-green, with a broad margin of pure white; habit quite dwarf; it is not affected in the least by exposure to the direct sunlight. Price, 15 cts.

Mrs. Parker. The foliage of this variety is identical with Mountain of Snow, but produces a quantity of beautiful double pink-colored flowers, which render it very distinct and attractive. Price, 25 cts. each.

GERANIUMS.

(Gold, Silver, and Bronze-Leaved.)

Alma. Leaves deep green, silver margin; flowers bright scarlet.

Bijou. Leaves round, silver margin; flowers bright scarlet.

Crystal Palace Gem. Leaves golden, green centre.

Countess of Warwick. Sulphur-white, carmine zone.

"   " **Craven.**

Emperor of Brazil. One of the best bronze varieties.

Freak of Nature. An improvement on "Happy Thought," foliage much smaller; habit very dwarf and branching; centre of leaf pure white.

Happy Thought. This is a novel and interesting variety, having a large yellow blotch in the centre of the leaf, with an outer band of green at the margin; flowers rich magenta-rose; habit, dwarf. '

Lady Cullum. Improvement on Mrs. Pollock. Price, 25 cts. each.

Marshal McMahon. Ground-color of the leaves golden yellow, deep chocolate zone, extra fine variety. Price, 25 cts. each.

Mountain of Snow. Deep green, white margin.

Mrs. Pollock. Leaves deep green, zone tinted with scarlet, yellowish green margin.

Mrs. J. Clutton. A fine silver tricolor.

Miss Goring. Golden yellow leaves, margined with brilliant flame-colored tints; crimson-scarlet zone. Price, 50 cts. each.

Prince Silver Wings. Green disk, bronze zone, tinted with carmine and white.

Prince of Tricolors. Bright green disk, with crimson zone, interspersed with fiery scarlet.

Socrates.

Wm. Sandy. Like Miss Goring in color, but of a stronger growth. Price, 50 cts. each.

Price, 15 cts. to 25 cts. each, except where noted; $1.50 to $2.25 per doz.

GERANIUMS.

(Hybrid Bedding and Scented.)

Lady Plymouth. Variegated foliage, rose-scented.

Morganii. Dark scarlet flowers.

Prince of Orange. Delicious scent.

Radula. Singular-cut foliage.

Rose.

Unique (Rollinson's). Deep crimson-purple flowers.

White Unique. Light flowers, spotted.

Price, 20 cts. each; $2.25 per doz.

GERANIUMS.

(Ivy-Leaved.)

Bridal Wreath. Pure white.

Lady Edith. Leaves green, flowers crimson, shaded purple.

L'Elegante. Foliage bright green, with broad bands of pure white; flowers pure white; fine for baskets, etc.

Willsii. A new hybrid of vigorous habit and compact growth, flowers violet-rose color.

" **Rosea.** Like the last in habit, with delicate rosy pink flowers.

Price, 15 cts. to 30 cts. each; $1.50 to $2.50 per doz.

DOUBLE IVY-LEAVED GERANIUMS.

Finette. Blush-white, upper petals flushed with rose and feathered with dark crimson.

Kœnig Albert. Violet-rose color.
Mme. E. Galle. Pure white, very double and large; the best.
M. Dilbus.
Viscountess Cranbrook. White, shaded satin-rose. Very pretty; quite
    double.

Price, 15 cts. to 20 cts. each; $1.50 to $2.00 per doz.

GRAPTOPHYLLUM NORTONII.

A foliage-plant of neat habit. The leaves are of ovate-oblong form, about six
inches long and two and a half broad; deep glossy green. The mid-ribs are at
first light rose, and have on one or both sides a central blotch of light yellow. As
the leaves approach maturity, the mid-ribs deepen to crimson, and the blotches
become suffuse with rose. Price, 50 cts. each.

HELIOTROPES.

A well-known and much-admired fragrant plant. Our collection contains the
best of the light and dark varieties. Price, 10 cts. to 30 cts. each; $1.00 to $3.00
per doz.

HERBACEOUS PLANTS.

Herbaceous plants are the most valuable of all garden flowers. They are not
only hardy, but easy of culture, increasing rapidly, and give a succession of bloom
from May to November.

We have added a large number of choice new and old varieties of this valuable
class of plants to our stock this year, that are not enumerated in the following
list : —

Anemone. In variety.
Aquilegia (Columbine). Well-known flower, throwing up stems of flowers of
    various colors, two feet high.
Baptesia. Handsome spikes of blue, lupine-shaped flowers, in June and July.
Daisy (Bellis). See special description.
Dielytra (Bleeding Heart). A beautiful border-plant, with brilliant rosy, heart-
    shaped flowers.
Delphinium. Four sorts.
Canterbury Bell. Large, showy, bell-shaped flowers of pure white, rose, and
    purple; June.
Fox Glove (Digitalis). Showy, bell-shaped flowers, on stems three or four feet
    high.
Flaxinella. Small spikes of white and reddish purple flowers, in June.
Funkia Japonica Alba. White.
    "   "   Foliage variegated, light blue.
Gladiolus Colvilli (The Bride). The finest white gladiolus among the early-
    flowering section; invaluable for market purposes, excellent for cut-flowers.
    It is quite hardy, and will grow in any light soil.
Helleborus Niger (the common Christmas rose). Greenish white.
    "   "   In variety.

Hyacinths. In variety.

Hollyhock. A fine collection of all colors.

Iris. In variety.

Iberis Sempervirens (Perennial Candytuft). Low-growing plants, with heads of white flowers.

Lychnis Viscaria. Pinkish red, double flowers.

Lily of the Valley. White.

Myosotis (Forget-me-not). Blue.

Narcissus Poeticus Ornatus. Flowers large, pure white, with a red cup.

Narcissus. In variety.

Phlox. (See special description.)

    "   **Subulata** (Moss Pink). Low-growing, covered in spring with pink flowers.

Polyanthus. In variety.

Primulas. In variety.

Spiræa Japonica. Feathery white flowers, one foot high.

    "   "   Fol. variegata.

    "   **Palmata.** The finest of the "Meadow Sweets"; large corymbs of crimson-purple flowers, three feet high.

    "   **Plumosa.**

    "   In variety.

Sedum. For rock-work. See special description.

Sweetwilliams. In variety.

Vinca (Periwinkle). White and blue, for cemetery.

Walhenbergia. Blue and white.

Yucca (Adam's Needles). Showy, cream-white flowers.

Price, 25 cts. to 50 cts. each ; $2.50 to $5.00 per doz.

HERNIARIA GLABRA.

A neat, low-growing plant, with dark, glossy-green leaves; a valuable plant for fancy bedding. Price, 15 cts. each; $1.50 per doz.

HIBISCUS.

A showy and handsome plant, of rapid growth, forming large bushes. The foliage is a glossy green. The Hibiscus is valuable both for summer and winter blooming.

H. Cooperii Tricolor. Foliage beautifully mottled with pure white and pale rose color; flowers bright crimson.

H. Double. Red.

    "   "   Crimson-scarlet.

Price, 25 cts. to 50 cts. each; $2.50 to $4.00 per doz.

HYDRANGEA SPECIOSA.

A very conspicuous and showy variety, having very large leaves, with a blotch of pure white through the centre of each. Price, 35 cts. each.

HYDRANGEA JAPONICA VARIEGATA.

Beautiful foliage of deep green, marked with white. Adapted to a rather shaded location in the garden, as the leaves burn when exposed to our hot mid-summer sun. Excellent for green-house decoration during the summer. Price, 25 cts. each; $2.50 per doz.

HYDRANGEA PANICULATA GRANDIFLORA.

A beautiful, hardy variety from Japan, with immense trusses of pure-white flowers, changing to blush, and remaining in blossom from August until frost. Price, 50 cts. each.

HYDRANGEA OTASKA.

A fine variety of this popular plant, producing immense pannicles of rosy carmine flowers when the plants are quite small. Price, 30 cts. each; $3.00 per doz.

HYDRANGEA.

(Thomas Hogg.)

One of the novelties introduced from Japan. It belongs to the Hortense section of the family, but it is a far more free and abundant bloomer than any other kind. The flowers are of the purest white, and of very firm texture. The plants blossom when quite small, and continue in bloom for a great length of time. A valuable plant for cemetery decoration. Having a large stock, we have put them at a very low price. Price, 25 cts. to $2.00 each.

IRIS.

The Iris are a very beautiful class of plants. They are of various shades of blue, purple, white, and yellow. Perfectly hardy, and require no attention after being planted. Price, 25 cts. each; $2.50 per doz.

IXORAS.

Beautiful stove-flowering plant. Producing large truss of orange-colored flowers.

I. Willliamsii.   I. Coccinea Superba.

JAPANESE MAPLES.

These Maples are among the most beautiful and interesting additions to our ornamental deciduous trees or shrubs, that have been made within the past few years. They are low trees or shrubs, with erect stems, the branches more or less spreading in the different varieties, and clothed with foliage that is developed in a greater variety of form and color than in any other species of deciduous trees known. Price, $1.00 each.

KLEINIA REPENS.

A very pretty succulent plant, with long, roundish, glaucous leaves. A very desirable basket-plant. Price, 25 cts. each.

LANTANA CALIFORNICA.

One of the finest varieties out; the habit is very dwarf and compact; the flowers are bright golden orange; an abundant bloomer, very desirable for bedding. Price, 10 cts. to 25 cts. each; $1.00 to $2.50 per doz.

LEUCOPHYTON BROWNII.

A very pretty variety of white or silvery-leaved plants, for ribbon or carpet bedding. Price, 10 cts. each; $1.00 per doz.

LOBELIAS.

Plants of easy growth, well adapted for bedding, edging, rockeries, baskets, etc. Price, 10 cts. each; $1.00 per doz.

LOMARIA GIBBA.

A green-house Tree-fern, of the most elegant growth, adapted for every purpose, graceful in its feathery fronds, easy of culture in any shady wood-earth; grows rapidly; adapted for baskets, vases, etc.; bears exposure to the sun like a Sago Palm, to which it has a strong resemblance. Price, 15 cts. to 50 cts. each; $1.50 to $4.00 per doz.

LYGODIUM SCANDENS.

(Japanese Climbing Fern.)

A climbing fern of graceful, twining habit, often attaining a height of twenty feet, and is much used as a substitute for smilax in decorating. It is of easy culture, and is a handsome plant for vases or hanging baskets, as it can be used to climb or to droop as required. Price, 25 cts. each; $2.50 per doz.

MARANTA.

A beautiful ornamental foliage plant, suitable for ferneries, baskets, vases, etc.

Argyrea.	Makoyana.	Regalis.
Bicolor.	Micans.	Van den Heckei.
Litziana.	Pulchella.	Zerina.

Price, from 25 cts. to $2.50 each.

MAHERNIA ODORATA.

A very desirable green-house plant, blooming profusely during the spring months. The flowers are yellow, bell-shaped, and very fragrant. Price, 25 cts. each

MYRSIPHYLLUM ASPARAGOIDES.

(Smilax.)

There is no plant in cultivation that surpasses this in the graceful beauty of its foliage, and its peculiar wavy formation renders it one of the most valuable of all plants for vases and baskets, as it can be used either to climb or to droop as required. The foliage is smooth and glossy; indispensable for bouquets, wreaths, etc. Price, 10 cts. each; $1.00 per doz.

MAURANDIAS.

Beautiful climbing plants, of graceful, slender growth, producing an endless profusion of handsome white, pink, or purple gloxinia-shaped flowers throughout the season.

Alba. White.   **Barclayana.** Purple.   **Rosea.** Rose.

Price, 10 cts. to 25 cts. each; $1.00 to $2.00 per doz.

NIEREMBERGIA.

Very beautiful, neat-growing plants, of good habit and delicate foliage, with a profusion of pretty purple and white flowers from June to September; well adapted for baskets, vases, etc.   Price, 10 cts. to 25 cts. each; $1.00 to $2.25 per doz.

ORCHIDS.

As the prices of Orchids vary considerably, according to the size of the plants, we think it best not to affix any price to this list; but we shall be pleased to give all information on application.

Calanthe Veitchii.
"   Vestata.
Cœlogne Cristata.
Cyprepediums in variety.
Dendrobium Bigibbum.
"   Bensonii.
"   Crassinode.
"   Crystallinum.
"   Densiflorum.
"   Devonianum.
"   Formosum.
"   "   Giganteum.
"   Nobilis.
"   Pierardi.
"   Thyrsiflorum.
"   Wardianum.
Epidendron Vitellinum.
Lælia Albida.

Lælia Anceps.
"   Autumnalis.
"   Purpurata.
Odontoglossum Alexandra.
"   Cirrhosum.
"   Grande.
"   Pulchellum.
"   Vexilarium.
Oncidium.
"   Forbesii.
"   Incurvum.
"   Tigrinum.
"   Varicosum.
Peresteria Elata.
Phajus Grandiflora.
Pilumina Fragrans.
Zygopetalum.

PANSIES.

We have a very choice collection, raised from the seed of the best English and American growers. Price, 10 cts. to 15 cts. each; $1.00 to $1.50 per doz.

PÆONIES.

One of the showiest and most desirable of hardy, ornamental plants, with beautiful, sweet-scented flowers. Their easy culture recommends them to every one who has a garden. Price, 25 cts. to 50 cts. each; $2.50 to $4.00 per doz.

PALMS.

As a decorative plant they stand unrivalled. They are easily cultivated, and require but little attention; of graceful and stately form, they impart a rich, sub-tropical appearance wherever used.

Areca Luteacens.
" **Rubra.**
" **Sapida.**
Cocos Weddelliana. A most elegant, small palm, with graceful, arching leaves, of a rich, dark green.
Cycas Revoluta. The sage-palm, very choice; $1.00 to $5.00 each.
Dioon Edule.
Kentia Belmoreana.
Latania Borbonica. Beautiful, with immense, fan-shaped leaves; $1.00 to $10.00 each.
Phœnix Dactylifera. Date-palm.
" " **Reclinata.**
Ptychosperma. A fine palm.

Price on application.

PASSIFLORA.

(Passion Vine.)

A beautiful, free-growing climber, with handsome, dark green leaves and showy flowers.

Passiflora Cærulea. Blue.
" **Decasiana.** Pale blue-red and purple.
" **Variegata.** Same color, with variegated leaves.

Price, 25 cts. each; $2.50 per doz.

PEPEROMIA.

Beautiful, low-growing plants, with variegated foliage. A choice plant for fern-eries, etc.

Maculosa. Price, 25 cts. each.   **Verschaffeltii.** Price, 25 cts. each.

SHOW AND FANCY PELARGONIUMS.

These Pelargoniums are remarkable for their large, showy forms, easily grown, flowering profusely, and present in their varied tints and colors what few other plants possess.

We mention below some of the best varieties for market and general decorative purposes. They embrace some of the finest varieties yet introduced.

Adanson. Deep purple-crimson, violet centre.
Alexina.   .
Beadsman.
Beauty of Oxton. Upper petals veined rich maroon, darkly blotched; under petals dark crimson, shaded with maroon; light centre, tinted with rose; all the petals are regularly margined with white, and beautifully fringed.

Beauty of Kingston. Pearly white, with a dark spot on each petal.

Blanche Fleur. Nearly white.

Braid's Decorator. Crimson and maroon-blotched, veined in lower petals; dwarf and free.

Charles Keen. White centre, rich lower petals, maroon top, shaded with carmine.

Coronet. Crimson-rose, dark crimson spots.

Corsair. Light purple, with pure white centre; petals blotched with maroon-crimson.

Desdemona. Pure white, deep maroon blotch.

Digby Grand. Blush-white, with veined blotch on the upper petals, fringed edge.

Dr. Andry.

Dr. Masters. Large, black blotches in the centre, margined with rich crimson, lower petals smaller blotch, with broader margin than the upper.

Duchess of Bedford. Pure white, slightly pink on top petals, and very flowering.

Duchess of Edinburgh. Delicate white, blotched with purple.

Grandes. Carmine, light centre, maroon spot on every petal, good habit, and free-flowering.

Harlequin. Carmine, dark spot, white throat.

Her Majesty.

Imperatrice Eugenie. French white, rich, dark spots.

John Bright.

Mabel. Dark maroon, narrow edge of carmine.

Maria.

Maria Mallet.

Mary Hoyle. Orange-rose, white centre.

Minnie.

Miss Bradshaw. Bright carmine spot, fringed edge; improvement on "Dr. Andry."

Mme. Kœnig.

Morning Star. Pure white, with a very deep crimson blotch in upper petals; flowers large; habit dwarf and spreading.

Mrs. Gladstone.

Mrs. L. Floyd. Bright scarlet, profuse-bloomer.

Le Negro.

Perseverance.

Prince of Pelargoniums. Vermilion-scarlet, blush-white centre, veined violet, upper petals flushed with crimson.

Purity. Large, pure white, maroon-carmine spot on upper petals.

Reine Hortense.

Robert Green.

Rose Celestial. Delicate pink, deep rose blotch.

Rosa Munda.

Scarlet Gem. Rosy scarlet, light centre, fringed, very beautiful.

Sir J. Paxton. Very dark purple, free-bloomer.

Triomphe de St. Mand. Rich, deep crimson, large trusses; dwarf, strong habit, free.

Vestal. Pure white, dark spots on upper petals.

Whitstone Herr.
Wm. Bull. Crimson-scarlet, dark spotted, free-bloomer, very fine.

Price, 25 cts. to 50 cts. each ; $3.00 to $5.00 per doz.

PILEA MUSCOSA AND SERPÆFOLIA.

Very pretty, neat-growing plant, with slender, frond-like leaves ; desirable for baskets, vases, etc. Price, 15 cts. each ; $1.50 per doz.

PETUNIAS.

(Single, Striped, and Blotched.)

Our stock of Petunias has been much improved by saving only from the best flowers, so that we now have a superior strain, which for beautiful markings and solidity of flowers can scarcely be equalled. Price, 10 cts. each ; 75 cts. per doz. ; $5.00 per hundred.

PHLOX.

A fine collection of this most interesting of our hardy, herbaceous, perennial plants. They are of the easiest culture, and embrace every color, from purest white to deepest crimson ; well adapted for city gardens or shrubberies, as they will grow either in shade or sunshine.

Advancer. Purplish crimson, dark eye.
Argosy. Crimson, centre bright, dark eye.
Augusta Riviere. Pure white, purple-crimson eye.
Augusta S. Riviere. Bright vermilion, crimson eye.
Brilliant. Dark rosy pink, large crimson eye.
Calypse. Violet-crimson, large.
Chas. Wagner. Deep, bright crimson, maroon eye.
Coccinea. Crimson-scarlet, fine.
Countess of Breadalbane. Dark, purple-crimson.
Cyprus. Salmon-pink, dark eye.
Defiance. Violet-crimson, bright.
Dr. Bourdevall. Beautiful carmine-rose, large.
Duke of Lancaster. Deep rose, dark centre.
Edith. Pure white, violet-purple eye.
Ernest Benary. Pale rose, deep eye.
Gloire de Neuilly. Deep vermilion, shaded crimson, dark eye.
J. K. Lord. Vermilion, shaded crimson, bright eye.
John Forbes. Clear pale rose, crimson eye, very large and fine.
Jeanne d'Arc. Snow-white, fine.
Larina. Pure white.
Louis Van Houtte. Vermilion, maroon eye.
Lothair. Deep crimson, shaded scarlet, carmine centre.
M. Dubue. Salmon-pink, crimson eye.
Maria Laison. Magenta, shaded violet, dark eye.
Mme. Forbois. Blush, large violet-crimson eye.
Mme. Brandtt. White, bright eye.
Mr. Ware. Purplish crimson.

Pomme de Sidon. Dark violet-crimson, fine.
Resplendent. Magenta, shaded scarlet, fine.
Ruby. Reddish purple, fine.
S. S. Ware. Dark lilac, light centre.
Splendens. Magenta-rose.
Souvenir. Bright vermilion, dark eye.
Virgo Maria. Fine white.
York and Lancaster. Light rose, purple-crimson eye.

Price, 25 cts. each; $2.50 per doz.

PANICUM VARIEGATUM.

A very pretty, neat-growing, variegated grass, of drooping habit. The color of the leaves is dark green, rose, and white; desirable for baskets, vases, etc. Price 15 cts. each; $1.50 per doz.

PANDANUS GRAMINIFOLIUS.

A fine, dwarf-growing variety, of drooping habit: with deep green, narrow leaves, serrated on the edges. Price, 50 cts. each.

PANDANUS UTILIS.

A beautiful plant, with long, recurved, glossy green leaves; well adapted for centres of vases, baskets, etc. Price, 50 cts. to $1.00 each.

PANDANUS VEITCHII.

A most beautiful variety, leaves light green in color, with stripes and bands of pure white; most graceful habit. Price, $1.00 to $5.00 each.

PHORMIUM TENAX VARIEGATUM.

A variegated form of the well-known New Zealand Flax. The leaves, which are often six feet long, are of a dark green color, distinctly marked with broad stripes of yellow, more than half the leaf being frequently of the latter color. It is very ornamental for conservatory and summer flower-garden decoration. Price, $3.00.

PERISTROPHE ANGUSTIFOLIA AUREA.

A beautiful plant, with brilliant golden foliage, striped with green; excellent for baskets and vases. The flowers are a delicate violet-color; free-bloomer. Price, 15 cts. each; $1.50 per doz.

POINSETTIA PULCHERRIMA.

A magnificent plant, producing, about Christmas, the most gorgeous bracts of rich vermilion, which often measure one foot in diameter, and remain in perfection several weeks. Price, 50 cts. to $1.00 each.

POINSETTIA PULCHERRIMA PLENISSIMA.

One of the most gorgeous of plants. The bracts, or flower leaves, are often over a foot in diameter, and of a most brilliant vermilion. Price, 50 cts. to $1.00 each.

REINECKIA CARNEA VARIEGATA.

A beautiful, variegated, grass-like plant of dwarf habit; an excellent plant for aquariums, ferneries, etc. Price, 25 cts. each.

RHYNCHOSPERMUM JASMINOIDES.

A beautiful green-house climber, flowers pure white and very fragrant; a valuable winter-blooming plant. Price, 25 cts. each.

RHODODENDRONS.

We would call the attention of our patrons to our superior stock of hardy Rhododendrons. These beautiful plants are yet but little cultivated, undoubtedly in a great degree from the general impression that they are difficult to manage, requiring special care and what is called peat soil; and it was at one time believed that they would not thrive in any other. Experience, however, proves the contrary, and it is now found that Rhododendrons thrive in almost any soil that does not contain lime. Strong plants, well set with buds, from two to three feet high, $2.00 to $5.00 each.

HYBRID TEA ROSES.

Camoens. A bright China rose.
Beauty of Stapleford. Pale pinkish rose, shaded darker in centre.
Duke of Connaught. Dark, rich crimson; immense size.
Hon. Geo. Bancroft. Bright rosy crimson, edged with bright red; large, and of fine form.
Jean Sisley. Rosy lilac, turning to bright pink in centre; large and full.
Nancy Lee. Bright satin rose, of good form.
Pierre Guillot. Bright dazzling red; petal bordered with white; very large, with erect flowers, well formed; very free bloomer; growth vigorous.
Reine Marie Henriette. A red Gloire de Dijon; large and full; climbing. This superb variety is the only one in this class.
Viscountess Falmouth. Very delicate pinkish rose, tinged with darker pink; large and of good form.

Price, 25 cts. to 50 cts. each; $2.50 to $4.50 per doz.

CHOICE TEA ROSES.

Catherine Merrmet. Delicate flesh-color; large, fine show-flower.
Comtesse Riza du Parc. Beautiful metallic rose, changing to pink; large, full, and good form.
Cornelia Cook. Creamy white; very large.
Duchess of Edinburgh. Crimson; most desirable color; fine form.

6

Louis de Savoie. Clear pale yellow; very large and full.
Mme. Lambard. Fine bright red; large, full, distinct, and beautiful.
Pearl des Jardins. Straw-color; very fine, one of the best.
   " de Lyons. Deep fawn and apricot; a beautiful large rose.
Souvenir d'un Ami. Salmon-rose; very large and full; a superb rose.

Price, 25 cts. to 50 cts. each; $2.50 to $4.50 per doz.

NOISETTE ROSES.

(Climbing Roses.)

Lamarque. Pure white, in clusters.
Marechal Niel. A rich brilliant yellow, too well known to need any description.
Mme. Caroline Kuster. Centre canary-yellow, outer petals pale lemon, flowers large, globular; a fine rose.

Price, 25 cts. to 50 cts. each.

MONTHLY ROSES.

Agrippina. Brilliant crimson.
Bon Silene. Rich, deep pink.
Douglass. Very deep maroon-crimson.
Malmaison. Blush, fine flower.

Niphetos. Pure white, extra.
Pauline Labonte. Light blush.
Safrano. Orange-yellow, splendid in bud, free-bloomer.
Yellow Tea.

Price, 15 cts. to 50 cts. each; $1.50 to $5.00 per doz.

HYBRID PERPETUAL ROSES.

Most of our roses are grown in pots, thereby avoiding all risk of transplanting and ensuring to the purchaser plants that are alive, and the certainty of their growth. This class of roses are entirely hardy, and bloom in spring and fall; they are greatly prized for their large, full-shaped flowers, which they produce in abundance.

Purchasers leaving the selection to us, will get a fine assortment of varieties and first-class plants. Price, 50 cts. to $1.00 each; $5.00 to $10.00 per doz.

PRAIRIE ROSES.

Baltimore Belle.
Elegans.

Gem of Prairies.
Queen of the Prairies

Price, 50 cts. each.

MOSS ROSES.

We keep only the best varieties. Price, 50 cts. to $1.00 each; $5.00 per doz.

SALVIAS.

Betheli. A splendid variety, excellent for autumn and winter blooming; dwarf habit, with beautiful pea-green foliage, and produces large spikes of the brightest rose.

Pitcherii. Another of the hidden genus; color most intense blue, flowers small in comparison with other varieties, more like a lobelia, but blooms very persistently, and set very close; for cut-flowers all winter, it is most valuable.

Splendens Bruanti. This is much the best of all the brilliant scarlets for autumn and winter blooming; habit dwarf, foliage pea-green; very free-flowering; spikes large and of a most intense color.

Price, 50 cts. each.

SONERILA HENDERSONII.

An exceedingly beautiful stove-plant, of dwarf habit. The leaves are an olive-green, densely marked with silvery white pearl-like spots. Price, 25 cts. each.

SONERILA HENDERSONII ARGENTEA.

Another handsome variety, with silvery leaves; in fact, the entire plant seems to be made of frosted silver. Price, 25 cts. each.

SUMMER AND OTHER CLIMBERS.

	Each.
Ampelopsis Veitchii. Small variety of the Virginia-Creeper .	$0 25
Bignonia Radicans	50
Chinese Wistaria	50
Clematis. In variety	50
Coboea Scandens	25
" " **Variegata**	50
Honeysuckle. Monthly, blooms all summer, very fragrant	25
" Scarlet trumpet, monthly, blooms all summer, very showy	25
Ivy. English	$0 25 to 1 00
" Variegated	35 to 1 00
Lonicera Aurea Reticulata. A beautiful, hardy climber, leaves being bright green, netted all over with yellow veins	25
Maurandia. Blue and white	15
Madeira Vine	15
Passion-Flower. In varieties	25
Tropæolum. In variety	10 to 15
Vinca Minor	15 to 50
" **Major.** Variegated	15 to 50
Virginia Creeper	25 to 50

SUMMER-FLOWERING BULBS.

Each.

Lilium Auratum, or Golden-banded Lily, universally acknowledged to be the finest of all lilies	.	$0 50
" **Candidum,** is the well-known white garden-lily, fragrant	. .	25
" **Harrisii.** New, white, fragrant	. .	75
" **Lancifolium Album.** Pure white	. .	50
" " **Rubrum.** White, spotted with crimson .	. .	50
" " **Roseum.** White, spotted with rose	. .	50
" **Longiflorum.** Long-tubed, white	.	25
Tritomas	.	25

Tuberose. One of the choicest summer-flowering bulbs; flowers are white and very fragrant; indispensable for making bouquets: —

Started in Pots, per dozen	3 00
Dry Roots " "	1 00

SANTOLINA CHAMÆCYPARISSUS.

One of the best plants for edgings or ribbon-lines, growing about a foot high. Price, 15 cts. each; $1.50 per doz.

SANCHEZIA NOBILIS VARIEGATA.

A very handsome hot-house plant, with leaves from twelve to fifteen inches long, beautifully veined, and marbled with deep, golden yellow. Price, 30 cts. to 50 cts. each.

SEDUM.

(Stone Crop.)

The Sedums belong to the same family as the Echeverias and Sempervivums. Being succulent plants, they are among the most valuable plants we have for hanging-baskets, vases, and rock-work. They are also extensively used for the edging of beds, or forming outer marginal lines, being dwarf and compact. The flowers of some of the varieties are very beautiful. They are of the easiest culture, either for garden or parlor decoration.

SEMPERVIVUMS.

(House Leek.)

A succulent genus of plants allied to the Sedums. They are unsurpassed for rock ornamentation, many of them being hardy.

SIBTHORPIA EUROPEA.

A neat little plant of creeping habit, very effective when grown in baskets, giving it a graceful appearance; an excellent plant for rock-work, moss-baskets, etc. Price, 10 cts. each; 75 cts. per doz.

STATICE HALFORDI.

One of the most beautiful green-house plants, with large, broad, light green foliage; flowers bright blue and white, very showy, and remain in perfection two or three months if kept in a cool house. Price, $1.00 to $5.00 each.

STEPHONOTIS FLORABUNDA.

A beautiful old vine, with deliciously fragrant flowers, which are borne in clusters similar to Bouvardia, but of much heavier and wax-like texture; fine for cut-flowers. Price, 25 cts. to 50 cts. each; $2.50 to $5.00 per doz.

TORENIA ASIATICA.

A very pretty summer-plant, with small blue flowers, gloxinia-like shaped; well adapted for summer-baskets, vases, etc. Price, 25 cts. each; $2.50 per doz.

TRITOMAS.

Splendid, half-hardy border-plants, flowering from July to October, and produce long spikes of orange-red and scarlet tubular flowers, each raceme from one to two feet in length. They are well-adapted for forming large, effective groups and beds, in which the numerous, terminal, flame-colored blossoms have a fine effect. Price, 25 cts. each; $2.50 per doz.

TUBEROSE.

One of the choicest summer-flowering bulbs; the flowers are white and very fragrant; indispensable for making bouquets. Price, started in pots, $3.00 per doz.; dry roots, $1.00 per doz.

TUBEROSE.

(Pearl.)

This variety is generally conceded superior to the old variety; the flowers are of double the size, and imbricated like a rose; of dwarf habit, growing only two feet in height. Price, dry roots, $1.00 per doz.; started in pots, $3.00 per doz.

VINCA VARIEGATA.

A beautiful fast-growing plant, with bright green leaves, edged with yellow, and flowers of deep blue. Well adapted for baskets, rock-work, or vases. Price, 15 cts. to 50 cts. each; $1.00 to $4.50 per doz.

SWEET-SCENTED VIOLET.

(Marie Louise.)

This is undoubtedly the best violet offered for years. It surpasses all other varieties in the profusion of its flowers. In color it is darker than the Neapolitan violet, double its size, and quite as fragrant. Price, 25 cts. each.

VERBENAS.
(General Collection.)

The following varieties of this popular plant we have selected from our large collection, the selections being made with great care, with a view to having the greatest variety of color combined with the best bedding qualities.

Admiral Corbet.

Aimee. Pale mauve, yellow eye.

Alba Perfecta. White, very fragrant.

Beauty of Oxford. Large, pink, fine.

Blue Champion. Indigo-blue.

Bolivia.

Canobie. Carmine, shaded violet, white eye.

Celestial Blue. Extra fine, blue.

Colossus. Crimson, violet eye.

Cuba. White, rose-pink stripe, distinct.

Dazzle. Blood-red, black markings.

Diana.
Faust. Rosy salmon, shading to pink.
Gambetta.
Honor. Blush, dark centre.
Jewel.
Jubilee.
London Pride. Large, claret-color.
Lolo. Pink-white eye.
Lulu.
Marco.
Mozart. Splashed scarlet and white.
Mrs. Woodruff. Bright scarlet —the best scarlet.
Negro. Black, extra fine.

Niobe. Large, pure white.
Profusion. Large, blush, extra.
Raccoon. Violet-purple.'
Rosea Alba. Light pink, white eye.
Rose Queen.
Rover. Rich, dark maroon.
Scarlet Circle. Dazzling scarlet, large, white eye.
Sylph. Pure white.
Silver Plume. Finest white.
Vulcan. Flame-color.
White Beauty. White, large, and fine.
Zenobia. Purple, large white eye.

Price, 75 cts. per doz.; $5.00 per hundred.

BASKET-PLANTS.

Achyranthus. In variety.
Acorus Gramineus. Variegated.
Alyssum. Variegated.
Alternantheras. Four varieties.
Balm. Silver.
Caladiums. Varieties.
Centaurea Gymnocarpa. Silver-gray foliage.
Cineraria Maritima.
Cissus Discolor.
Coleus. Varieties.
Coronilla Glauca. Variegated.
Cuphea. Cigar-plant.
Dracænas. Red and green.
Euonymus. Variegated.
Ferns. In variety.
Ficus Repens.
German Ivy.
Gnaphalium Lanatum. Downy white foliage.

Ivies. Of sorts.
Lineria.
Lobelias. In variety.
Lonicera. Variegated Chinese honey-suckle.
Lycopodium. Of sorts.
Maurandias. Two varieties.
Panicum Variegatum. A beautiful variegated grass.
Pelargoniums. Ivy-leaved.
Peperomia Maculosa.
Saxifraga Sarmantosa.
Sedum Carneum Variegatum.
Torenia Asiatica.
Tradescantia. Two varieties.
Tropæolum. Of sorts.
Vinca Major Variegata.
    "   Minor.

Price, 10 cts. to 25 cts. each; $1.00 to $2.50 per doz.

MISCELLANEOUS BEDDING-PLANTS.

	Per doz.
Achyranthus. In variety . . .	. $ 75
Ageratum Mexicanum. Best variety	. 1 00
Aloysia. Lemon Verbena . . .	. 1 50
Alternanthera. Four sorts . .	. 75
Alyssum. Variegated-leaved . .	. 75
Antirrhinums. In variety . .	. 1 00
Asters. A fine assortment . .	. 50
Balsams. Camelia-flowered .	. 50
Basket Plants. In variety .	1 50

Bouvardias. In variety	2	50
Calceolarias. In variety	1	50
Carnations. Monthly	1	00
Centaureas. Silver foliage		75
Chrysanthemums. Fifty sorts	1	50
Cineraria Maritima. Fine silver-leaved plant .		75
Coleus. In variety		75
Dahlias. Fifty sorts	2	00
Daisy. Spotted leaves	1	00
Daisies. Six sorts	1	00
Feverfew		75
" Golden feather		65
Fuchsias. Best bedding varieties . . .	2	50
Geraniums. All varieties	2	00
Gnaphalium Lanatum. Downy white foliage	1	25
Heliotrope. Best bedding varieties . . .	1	00
Lantanas. In variety	1	50
Lobelia. In variety		75
Madeira Vine		50
Maurandias. Two varieties	1	25
Nierembergia. In variety	1	25
Pansies		50
" Select seedlings	1	00
Petunias. Double	2	00
" Fine, single, blotched . . .	1	00
Phlox. Fine variety	2	50
Salvia Splendens	1	00
Stocks. Bedding	1	00
Tropæolums. Varieties		50
Verbenas. With names		75
Zinnia. Extra, double		50

SHRUBS.

Almond. Althæas. In variety.   **Azaleas.** In variety.   **Calycanthus.**
Corchrous. Silver-variegated leaved.
Deutzia Crenata Flora Plena. Flowers double, white, tinged with rose.
  "   **Gracilis.** Flowers pure white.
  "   **Scabra.** Profuse white flowers.
Forsythia Viridissima. Flowers bright yellow, very early in spring.
Fringe Tree.
Hawthorn. Double, pink and white.
Honeysuckle Tartarian. Red and white.
  "   **Scarlet Trumpet.** Monthly, blooms all summer, very showy.
  "   **Yellow Trumpet.** Very fragrant.
Lilac. Common purple, common white, and Persian.
Magnolia Tripetala. Fine, large leaves, white flowers.
Mahonia Aquifolia. Flowers yellow, early spring.
Purple Fringe, or, Smoke Tree.
Pyrus Japonica. Flowers bright scarlet, early spring.
Rhododendrons. In variety; $1.50 to $3.00 each.

Spiræa Prunifolia. Flowers white, blossoms in May.
" **Reevesii.** Round clusters of white flowers bloom in May.
Syringa, or, Mock Orange. Three varieties.
Tree Box.
Viburnum.
Weigelia Alba. Flowers white, changing to a light, delicate blush.
" **Rosea.** Rose-colored flowers, blooms in May.
" **Variegated-leaved.** Bordered with yellowish white, flowers pink.
Wistaria Sinensis. (Chinese.)
Price, subject to variety, 25 cts. to $1.00 each.

WEEPING OR DROOPING DECIDUOUS TREES.

Ash. European weeping; the common well-known sort; one of the finest lawn and arbor trees.
" Mountain weeping; a beautiful French variety, of rapid growth and decidedly pendulous.
Beech. Weeping; a very graceful tree.
Birch. Cut-leaved, weeping.
" European weeping.
Cherry. Weeping.
Elm. Weeping; an English variety, with smooth, glossy, leaves.
Poplar. Weeping.
Willow. American weeping.
" Kilmarnock weeping; one of the finest of this class of trees.
" Rosemary-leaved; branches feathery, with silver foliage.
Price, $1.00 to $5.00 each.

SHADE TREES.

A superior stock of Shade Trees of the kinds herein mentioned: Rock Maple, Horse Chestnut, Elm, Linden, Silver Maple, English Elm, Dwarf Horse Chestnut, and Magnolias. The Maples are unusually fine. Price, 50 cts. to $3.00 each.

FRUIT TREES.
(All the Leading Kinds.)

Pears. Standard, 75 cts. to $1.00 each. **Dwarf Pears.** 50 cts. to 75 cts. each.
Peaches. 25 cts. to 50 cts. each.

EXOTIC GRAPES FOR VINERIES.

All the leading kinds. Price, one-year-old plants, 50 cts.; two-year-old plants, $1.50 each.

SMALL FRUITS.

Currants. Best varieties, 15 cts. each; $1.50 per dozen.
Hardy Grapes. All the leading kinds. Strong plants, one-year-old, 25 cts. to $1.00 each; two-year-old, and extra vines, 50 cts. to $2.00 each.
Raspberries. 15 cts. each; $1.50 per dozen.
Strawberries.

HEDGE PLANTS.
(Evergreen, Stone.)

The idea of planting hedges for use and ornament, and screens for the protection of orchards, farms, and gardens, is a practical one, and rapidly becoming appreci-

7

ated. Among the trees adapted to ornamental hedges the American Arbor Vitæ and the Norway Spruce take the first place. By using medium-sized plants a hedge can be made as cheaply as a good board fence can be built, and then, with very little care, it is becoming every year more and more "a thing of beauty." We all know that such hedges constitute the principal attraction in our best-kept places.

American Arbor Vitæ. $8.00 to $25.00 per hundred.
Norway Spruce. $25.00 per hundred.

DECIDUOUS.

Privet. One-year, $5.00 per hundred.   **Japan Quince.** $15.00 per hundred.
Dwarf Box. For edging.

CHOICE CONIFERS.

Arbor Vitæ. Siberian, the best of all the genus for this country, exceedingly hardy, keeping color well in winter, growth compact and pyramidal, makes an elegant tree; 50 cts. to $1.00 each.

Arbor Vitæ. American Golden; beautifully marked with golden yellow, very hardy; $1.00 to $3.00 each.

Arbor Vitæ. Reidii, a dwarf bush of a beautiful shade of green, very dense and perfect form; 50 cts. to $2.00 each.

Biota Elegantissima. End of young branches tipped with golden yellow; $1.50 each.

Juniper, Irish. A tapering, pretty tree, $1.00.

  " **Swedish.** A small-sized, handsome, pyramidal tree, with bluish green foliage ; $1.00.

  " **Chinensis.** A small tree or shrub, with spreading]branches ; $1.00.

  " **Commonis.** A handsome, compact, small tree.

Pines, Austrian. A remarkably robust, hardy tree, growth rapid.

  " **Nordmanniana.** This is a symmetrical and imposing tree; $2.00.

  " **Pichta.** A medium-sized tree, quite compact and conical, and bearing very rich, dark foliage; $2.00.

  " **Cembra.** A handsome and distinct European species, of a compact, conical form, foliage short and silvery ; $1.00.

Retinospora Obtusa. A new evergreen tree from Japan, with beautiful, light green foliage; $1.00.

Retinospora Pisifera. A small tree, with numerous delicate branches and feathery foliage ; $1.00.

Retinospora Plumosa. A variety with fine, short branches and smalll leaves; $1.00.

Retinospora Plumosa Aurea Variegata. A beautiful variety, similar in habit to Plumosa, with foliage of rich golden-yellow ; small plants, 50 cts. each ; large plants, $2.00.

KITCHEN GARDEN ROOTS, PLANTS, ETC.

Asparagus. $1.00 per hundred.   **Rhubarb.** 25 cts. each ; $2.50 per doz.

PLANTS

Of Cabbage, Celery, Cauliflower, Egg Plant, Peppers, and Tomatoes, can be had in May or June.

INDEX.

Cut Flower Department.

———— • ————

THIS department is made a specialty, under the direct supervision of E. SHEPPARD & SONS, who have made every arrangement to have on hand a constant supply of the CHOICEST FLOWERS at all times, thus enabling them to execute orders received by mail or telegraph at short notice.

Bouquets, Baskets, Crowns, Crosses, Anchors, Wreaths, Masonic and other Emblems and Designs constantly on hand and arranged at short notice.

Designs can be packed to be shipped to a reasonable distance with perfect safety.

Orders received at C. R. KIMBALL'S Drug Store, corner Central and Merrimack Streets.

E. SHEPPARD & SONS,

☞ Telephone.   224 Fairmount Street, Lowell, Mass.

www.ingramcontent.com/pod-product-compliance
Lightning Source LLC
Chambersburg PA
CBHW022016190326
41519CB00010B/1537